POWELL'S C

by
W. L. Rusho

WITH ORIGINAL ENGRAVINGS AND PHOTOGRAPHS

FILTER PRESS

Palmer Lake, Colorado

1969

FILTER PRESS
Wild and Woolly West BOOKS

Phone 719-481-2523

P.O. Box 5, Palmer Lake, Colorado 80133

1. Choda — Thirty Pound Rails, 1956
2. Clemens — Celebrated Jumping Frog, 1965
3. Banks — Uncle Jim's Book of Pancakes, 1967, 1979
4. Service — Yukon Poems, 1967
5. Cushing — My Adventures in Zuni, 1967
6. Englert — Oliver Perry Wiggins, 1968 *(Out of Print)*
7. Matthews — Navajo Weavers & Silversmiths, 1968
8. Campbell — West Plates & Dry Gulches, 1970
9. Banks — Alferd Packer's Wilderness Cookbook, 1969
10. Faulk — Simple Methods of Mining Gold, 1969, 1981
11. Rusho — Powell's Canyon Voyage, 1969
12. Hinckley — Transcontinental Rails, 1969
13. Young — The Grand Canyon, 1969
14. Gehm — Nevada's Yesterdays, 1970 *(Out of Print)*
15. Seig — Tobacco, Peace Pipes, & Indians, 1971
16. Conrotto — Game Cookery Recipes, 1971 *(Out of Print)*
17. Scanland — Life of Pat F. Garrett, 1971
18. Hunt — High Country Ghost Town Poems, 1962, 1971
19. Arpad — Buffalo Bill's Wild West, 1971
20. Wheeler — Deadwood Dick's Leadville Lay, 1971 *(Out of Print)*
21. Powell — The Hopi Villages, 1972
22. Bathke — The West in Postage Stamps, 1973 *(Out of Print)*
23. Hesse — Southwestern Indian Recipe Book, 1973
24. Vangen — Indian Weapons, 1972 *(Out of Print)*
25. MacDonald — Cockeyed Charley Parkhurst, 1973 *(Out of Print)*
26. Schwatka — Among the Apaches, 1974
27. Bourke — General Crook in the Indian Country *and*
    Remington — A Scout with the Buffalo Soldiers, 1974
28. Powell — An Overland Trip to the Grand Canyon, 1974
29. Harte — Luck of Roaring Camp & other sketches, 1975
30. Remington — On the Apache Reservations & Among the Cheyennes, 1974
31. Ferrin — Many Moons Ago, 1976 *(Out of Print)*
32. Kirby — Saga of Butch Cassidy, 1977 *(Out of Print)*
33. Isom — Fox Grapes, Cherokee Verse, 1977 *(Out of Print)*
34. Bryan — Navajo Native Dyes, 1978
35. deBaca — Vicente Silva, Terror of Las Vegas, 1978
36. Underhill — Pueblo Crafts, 1979
37. Underhill — Papago & Pima Indians of Arizona, 1979
38. Riker — Colorado Ghost Towns & Mining Camps, 1979 *(Out of Print)*
39. Bennett — Genuine Navajo Rug; How to Tell, 1979
40. Duran — Blonde Chicana Bride's Mexican Cookbook, <u>1981</u>
41. Kennard — Field Mouse Goes to War, 1977
42. Keasey — Gadsden's Silent Observers, 1974
43. Beshoar — Violet Soup, 1982 *(Out of Print)*
44. Underhill — People of the Crimson Evening, 1982
45. Choda — West on Wood, 5 vols. *(In preparation)*
46. Duran — Mexican Recipe Shortcuts, 1983
47. Roosevelt — Frontier Types in Cowboy Land, 1988
48. Young — Kokopelli, 1990

ISBN 0-910584-12-5 paper

Copyright © 1969

PRINTED IN THE UNITED STATES OF AMERICA

ILLUSTRATIONS

The Gate of Lodore Canyon...............Front Cover
Camp at Flaming Gorge, 1871 Expedition...Title Page
John Wesley Powell, photo about 1879.............iv
Sentinal Rock, Wahweap Creek....................vii
Island Monument, Glen Canyon...................viii
Emma Dean II, 1871-1872 Expedition.................3
The start from Green River Station.................6
Ready to leave, 1871 Expedition....................7
The Gate of Lodore.................................9
Running a rapid...................................10
The wreck at Disaster Falls.......................12
Steamboat Rock in Echo Park.......................14
Echo Rock...15
Powell with the boats in Glen Canyon..............16
Light House Rock in Desolation Canyon.............18
Gunnison's Butte at mouth of Gray Canyon..........19
Narrow Canyon and Henry Mountains.................21
Cataract Canyon...................................24
Land of the Standing Rocks........................25
Noonday Rest in Marble Canyon.....................27
Camp at the Mouth of the Little Colorado..........29
Marble Canyon from the Vermilion Cliffs...........30
Marble Canyon from the Air........................31
In the Depths of the Grand Canyon.................33
Grand Canyon looking West from Toroweap...........35
The Grand Canyon viewed East from Toroweap........37
The Vulcan Rapid..................................38
Memorial Plaque at Separation Canyon..............40
Jacob Hamblin and Powell at Shivwits Council......43

Photographs courtesy of the U.S. Bureau of Reclamation.
Wood engravings are from the Filter Press Collection.

JOHN WESLEY POWELL 1834-1902

INTRODUCTION

With his spectacular conquest of the Colorado River, Major John Wesley Powell burst upon the American scene like a rocket shot across the Western sky. Acclaimed as a hero, Powell rode the tide of popularity into a commanding position in Government science, from which he was able to exert substantial influence on the American frontier.

Major Powell was a man who saw with clear insight into the tangled pattern of Western development. His genius lay in an ability to grasp the intellectual challenges of geology, ethnology, physics, chemistry, and civil engineering as they applied to the frontier, and out of these studies to forsee the problems that lay ahead for the settlers. Many hopeful pioneers, for instance, believed that the cultivation of land would induce more rainfall, the old "rain follows the plow" concept, but Powell riddled their ideas and called them foolish fantasies. He saw that the West was extremely varied in its topography and in its resources, and he said that the West should **not** be treated as if it were uniform farm land.

Unfortunately, Powell's was but a voice among many competitors. Since his proposals called for reasonable sacrifice and forebearance, his arguments were highly unpopular. The 1870's and 1880's were part of an era of wishful thinking so far as the West was concerned, a time when men believed what they wanted to believe, a time when a pot of gold lay at the end of every rainbow. Those earlybirds who had appropriated the resources of the West — particularly its water — were violently opposed to Powell.

In spite of the opposition, Powell's persuasive personality and driving intellect pushed ahead programs that led to the creation of the U.S. Geological Survey (of which Powell was the second Director), and later the Bureau of American Ethnology and the Bureau of Mines. For a while Powell headed the Irrigation Survey, an agency designed to map and classify the water resources of the West, but the agency was killed by the speculators and promoters who did not want people to wait for "facts" about the land before they settled on it.

Just before Powell died in 1902 he was told that the President, Theodore Roosevelt, had urged Congress to create what was to become the Bureau of Reclamation. It would mean that the Federal Government would participate in irrigation projects of the West.

Powell reflected a moment and then said, "These things take time. You must learn to control impatience, but always be impatient."

Before Powell became a hard-driving, impatient Bureaucrat, he had many adventures that molded the man and gave him direction. Some of the highlights include his teenage boat trips on the Ohio and Mississippi Rivers, his service as one of General Grant's officers in the Civil War, and his awakening interest in natural science that led to an appointment as a professor at Illinois Wesleyan College.

Nothing in Powell's earlier years, however, can compare in importance with his exploration of the Green and Colorado Rivers in 1869 and again in 1871-72. His idea for a boat trip down these heretofore unexplored rivers was not in any sense one of Powell's long time pet projects, but rather grew out of his field trips to Colorado Territory in 1867 and 1868.

Although Powell disclaimed all non-scientific motives for making the river trip, it was early recognized as one of the great adventures of all time. He and his men entered on an undertaking that most men thought doomed to failure, in which the crew would either be drowned or would have to climb out of the canyons to face probable death at the hands of the Indians or the desert. The fact that three of Powell's men were indeed killed only added to the miracle that Powell and five others escaped unharmed.

Powell's national fame was aided greatly by his stirring narrative account that was published by the Smithsonian Institution in 1875. Unlike most Government reports, Powell's read like a gripping novel. Using few unnecessary adjectives, his account is more like contemporary writing than most of the Victorian prose of his day. Perhaps this is one reason why his report has continued to be read by successive generations.

Detractors point out that Powell sometimes altered the actual history of the 1869 expedition by combining with it events that did not happen until the second expedition, or by changing the sequence of events. But such changes are minor to one who reads the report as a tale of adventure. Powell thought of the two trips as one exploration in two parts. Omission of the names of the personnel of the second expedition was perhaps justified by the dictates of good story-telling.

Complementing Powell's excellent report were a large number of woodcut drawings, which are actually artistic elaborations of photographs taken on the second expedition. The artist, usually Thomas Moran, altered some of the scenes to conform more closely with the events of the 1869 expedition. To some, Moran added artistic composition. Some of the woodcuts have been used as illustrations for this publication. They are used with full knowledge that actual, but inferior, photographs could have been substituted. If the exaggerated canyon walls sometimes makes the Colorado look like the River Styx, the reader must realize that it probably seemed so to the haggard boatmen.

Sentinal Rock, Wahweap Creek.

Photographer E.O. Beaman made his picture of Sentinal Rock in Wahweap Canyon for the 1871 voyage. The famous Western artist, Thomas Moran, dramatized it for this ISLAND MONUMENT wood engraving published in CANYONS OF THE COLORADO.

I
THE ROAD TO GREEN RIVER CITY

The civil War was over. A massive job of reconstruction awaited in the South, but the Nation could afford to turn much of its attention to the frontier West. Already settled were the Pacific Coast, the Rio Grande Valley of New Mexico, and the Great Salt Lake Valley. Although a few crude settlements, such as Denver, lay in or near the mountains, most of a wide region from the Rockies to the Sierras had been leapfrogged by the pioneers. Except for a few wagon routes, much of the intermountain area was classified as **terra incognita,** and was designated by large blanks on the map.

Major John Wesley Powell was discharged from the Union Army in 1865, his right arm gone, having been shattered by a minie ball at Shiloh, but his mind was sharper than ever. Before the war Wes Powell had been a public school principal in Hennepin, Illinois, where he had emphasized the study of natural science. Now, with the war over, he resolved to go more intensively into science studies. He therefore accepted a professorship at Illinois Wesleyan College, where he immediately became popular with the students.

Like many other men of his time, Powell was attracted by the mysteries of the American West, but as a scientist, Powell viewed the West as a vast outdoor laboratory that brimmed with marvels yet unknown.

In 1867, Powell led eleven amateur naturalists and students on a summer expedition to the Colorado Rockies. From Denver (Then a bustling city of 5,000), Powell's group headed into the mountains to the west and south, moving slowly so as to collect specimens of flora and fauna. Before returning they ascended Pikes Peak, which then lacked any sort of trail. Powell's wife, Emma, was the first woman ever to climb to the famous summit.

To Powell, probably the most interesting aspect of the mountains were the rivers, particularly the Grand, as the upper 400 miles of the Colorado were then called. Through September, 1867, Powell remained in the Rockies to examine the topography of the Grand while the remainder of his expedition returned to Illinois. His idea of exploring this river system by boat undoubtedly came to him during this first field trip to the West. Perhaps a river voyage was suggested to him by Jack Sumner, whom Powell had hired as guide. At any rate, it was Powell's driving force that eventually brought the idea to fruition.

By the summer season of 1868, nothing short of brute force could have kept Powell from making another field trip to Colorado. To finance it adequately, he visited his old commander, General Grant, in Washington, and asked for rations for 25 men for the season. In his written request, he made first mention of the canyons of the Colorado.

"It is believed that the Grand Canyon of the Colorado will give the best geological section on the continent..."

Finding that the Army had no authority to grant rations to civilians, Powell took his appeal to Congress, which eventually acceded to his request. During this period Powell also obtained instruments from the Smithsonian Institution.

Returning to Bloomington, Powell lined up his new expedition, which this time consisted of twenty-one persons. They headed west on the railroad to Cheyenne, where they obtained horses and pack mules for the trip south.

Prime purpose of the 1868 trip was to explore the high Colorado Rockies northwest of Denver. In late August, for instance, they became among the first men who had reached the summit of Longs Peak. The second objective of the expedition was to learn more about the Colorado River system. Since he had already investigated the Grand River, Powell resolved to have a look at the Green River and its tributaries. After most of the party returned to Illinois, Powell and a few associates moved into winter camp on the White River, a westward-flowing tributary of the Green. From there, Powell traveled north into Wyoming, visited Brown's Park, and looked into various Green and Yampa River canyons. He met some poor Ute Indians who struck another responsive chord in Powell's scientific curiosity. Investigating Indians however, could wait for another year.

Recent completion of the transcontinental railroad to Green River City meant that to this point — and to this point only — Powell could easily transport boats and other equipment needed for a river voyage. His decision was reached. His first exploration by boat of the Colorado River system would begin at Green River City and would terminate below the Grand Canyon. The time to embark would be during high water, or late May, of 1869.

As the year 1869 began, Major Powell was back in Illinois, busy with preparations for the trip. In Chicago he had four oak boats specially constructed to withstand the onslaught of waves and pounding against rocks. Airtight compartments in the ends of each boat assured buoyancy even when the open sections were filled with water.

For the all-important crewmen, Powell decided against us-

The Emma Dean II was used as Powell's flagship on the 1871-1872 Expedition. No chair was lashed to the deck of his boat in 1869. As the earlier expedition did not carry photographic equipment, all illustrations were from photographs of the 1871-1872 expedition.

ing scientists, however appropriate this might have seemed. Instead, he selected rugged outdoorsmen who could live off the land, and, if necessary, who would not flinch at incessant hard labor. Jack Sumner, for instance, was a professional guide and hunter. George Y. Bradley was an Army Sergeant until Powell asked President Grant to give him a discharge. William Rhodes Hawkins was a former Union soldier who just happened to be at Jack Sumner's trading post when Powell arrived in 1868. The backgrounds of William Dunn and Frank Goodman remain unknown to this day, although Goodman had apparently worked at the Uinta Indian Agency. Andy Hall, the cook, was a young professional hunter. Oramel G. Howland was a newspaperman working in Denver when he met Powell. Oramel's brother, Seneca, had lately arrived from Vermont when he was also asked to join the expedition. In addition to all of these, Major Powell brought along his younger brother, Walter. Although Walter Powell performed adequate service on the 1869 trip, he had not been emotionally stable since his mental breakdown suffered as a prisoner in a Confederate POW camp and was much inclined to moodiness.

In Washington to make final preparations for the trip, Major Powell again was able to secure an order from Congress and from the President that allowed him to draw Army rations for his men. But Congress refused to give any money. The Smithsonian gave equipment, Illinois State Normal and Illinois Industrial Colleges each gave five hundred dollars. A few of Powell's friends donated small sums of money. The Union Pacific Railroad issued passes for the men and their equipment for the trip to Green River City.

Arriving at Green River City on May 11th, Major Powell and his brother found the eight crewmen waiting for them. In those days the town consisted of little more than saloons and vacant buildings used a year earlier when the railroad was under construction. But in 1869 it was a dull place to sit around waiting.

Activity finally began when the four heavy boats were lowered from a flatcar and were launched into the nearby Green River. After camp was made along the river bank, the Major began drilling the men in boating procedure, in hand signals, and in elementary use of barometers. O. G. Howland was placed in charge of drawing a map as they traveled. All men had to share equally in the boating tasks.

On May 23rd Major Powell announced that all was in readiness and that the expedition would begin in the morning. Thereupon all hands adjourned to the local restaurant for one last meal in civilization.

II
A ROUGH BEGINNING

Townspeople of Green River City waved from the hill as the four boats pulled into the current and floated effortlessly down the placid water. The time was 1 p.m., May 24, 1869. Powell led in the lightest boat, the **Emma Dean,** which was named for the Major's wife. Following came **Kitty Clyde's Sister** and the **Maid-of-the-Canyon.** Bringing up the rear was the **No-Name.** Soon the boats rounded a bend and disappeared behind a low hill. The fate of the men now depended on skill, proper equipment and the river.

Fortunately, the first 50 miles were easy boating, for the frontiersmen that made up Powell's crew were inept boatmen, to say the least. They missed the Major's signals, their rowing was awkward, they misread the river, and they frequently grounded on sandbars. But the rough spots in their technique were rapidly worn off. By the time they reached the first canyon, on May 27th, they could at least keep the boats in the main current.

Powell described the first canyon with the following:

> At a distance of from one to twenty miles a brilliant red gorge is seen, the red being surrounded by broad bands of mottled buff and gray at the summit of the cliffs, and curving down to the water's edge on the nearer slope of the mountain. This is where the river enters the mountain range — the head of the first canyon we are to explore .... We have named it "Flaming Gorge."

At the Flaming Gorge they paused for three days while Powell mapped nearby Henrys Fork Valley and prepared a geological section.

Resuming their journey, they boated through the Flaming Gorge and were soon in the grip of their first mild rapids. They were able to run some rapids, but others required that the boats be "lined" down with ropes. The worst rapids required an actual portage of the boats — a tedious, time consuming procedure. Toward late afternoon on June first they reached a place where the river split around a great fallen block of sandstone. On a rock just above this rapid the men found a painted name and a date, "Ashley 1825."

Although Powell had no way of knowing it, he was not the first man to lead a party of boatmen through these upper canyons of the Green. In 1825, General William H. Ashley, just then beginning his fur trapping business, took some men through in frame boats covered with buffalo hide. Powell also knew

The departure from Green River City is dramatized by the artist working from photographs.

Actual equipment leaving in 1871. Powell (with one arm) is standing on boat at right. Note the large camera and attached field dark room at left.

absolutely nothing about a man named William Manly, who took a small group of California-bound prospectors down these canyons in 1849. Manly first used a converted ferryboat, and when that wrecked, he improvised crude dugouts. Both Ashley and Manly successfully negotiated their way through Flaming Gorge, Red Canyon, Lodore Canyon, Whirlpool Canyon, and Split Mountain Canyon.

The next day the Powell party reached a wide spot in the canyon known then as Little Brown's Hole, where they camped beneath some tall pines. Today the same spot is known as Little Hole, and is a highly popular place to catch the rainbow trout that thrive in the cold water flowing from Flaming Gorge Dam.

A few more miles and they were in Brown's Hole, which Powell renamed "Browns Park." Browns Park had been a favorite winter refuge for the mountain men Some years after Powell's voyage it became a refuge for outlaws, including Butch Cassidy and the Wild Bunch. But in 1869 it meant only a long, sluggish river meandering between the bluffs, where most progress was measured in laborious oar strokes. Here Powell stopped again for a few days to climb nearby mountains and to prepare a rough map.

On June 5th, as they floated on downstream, a canyon appeared in the distant mountain wall to the south, where the river was seen to glide into its dark and foreboding depths. Although their confidence remained firm, everyone on board felt a bit uneasy at this ominous sight.

That night in camp, Powell wrote:

Now, as I write, the sun is going down and the shadows are setting in the canyon. The vermilion gleams and the rosy hues, the green and gray tints are changing to sombre brown above, and black shadows below. Now 'tis a black portal to a region of gloom.

And that is the gateway through which we enter our voyage of exploration tomorrow — and what shall we find?

Powell may have had a premonition, for their first day in Lodore Canyon was one of the most disastrous of the entire trip.

After they entered the "Gate of Lodore" (as Powell named the opening to the canyon), they passed through a series of rapids, each, it seemed, worse than the ones preceding. Shortly after lunch, the Major, in the lead, came to a rapid he judged should not be run. But only two of the three other boat crews saw Powell's signal to land. The Howland brothers and Frank

The Gate of Lodore

Goodman, in the **No-Name,** kept on their course at "railroad speed." On shore the Major heard a shout and looked up to see the **No-Name** slide into the first rapid and veer out of control. A few hundred yards below was a much larger drop, where the river was divided by a large, nearly submerged, midstream boulder.

Running a rapid

Into the second rapid came the **No-Name,** so full of water by now that the men could not row effectively. The boat struck the midstream rock, throwing the men overboard. Just as the men were about to climb back on board, the craft struck another rock and split in two. The two Howlands and Goodman managed to reach a small island, around which the water raced. On the shore, Jack Sumner saw the emergency. Sumner quickly unloaded the **Emma Dean,** jumped in the boat and rowed swiftly to the island, where the three castaways climbed aboard. Some frantic oar work was necessary to get all four back to shore before the boat slid into the next rapid below.

One third of their rations had been lost, plus the clothing and gear belonging to the **No-Name's** crew. More important, all of the barometers, which they needed to measure their altitude and their descent down the river, were gone. Powell considered postponing the voyage until more barometers could be obtained.

Next morning, the Major examined the river from shore and discovered that part of the **No-Name** remained lodged on a rocky bar. Using the other boats to reach the wreck, the men searched for the precious instruments. When Powell heard a shout of triumph he felt sure that the barometers had been found. As it turned out the barometers had indeed been located, but the shout was not for them, it was for a two-gallon keg of whiskey that the Major did not even know was aboard the **No-Name.**

Lodore Canyon was named for a Robert Southey poem, "The Cataract of Lodore," which begins, "How does the water come down at Lodore?" Among other ways, Southey said, it comes down "sparkling, frothing, raging, leaping, twisting, fighting, hissing, roaring, battering and shattering, toiling and boiling." So the Green River, Powell observed, moves through this scenic gorge in the Uinta Mountains.

Strenuous labor saw them through the rest of Lodore. When they were not portaging equipment, they were lining boats around the rocks, or they were making camp, or trying to repair a damaged item of equipment. Frequently they saw small groups of mountain sheep, but when they stopped the "hunters" among the crew came down with buck fever and they only scared the game away. Bradley commented that "if left to maintain themselves with their rifles they would fare worse than Job's turkey."

One evening as they were unloading the boats, and as the cook was building a campfire among some rocks, a sudden gust of wind whipped sparks into the dry grass and brush. The whole beach quickly burst into flame. Since the men could not

The wreck at Disaster Falls

scale a sheer cliff, their only escape was the boats. But near tragedy was turned into pure comedy as the men came running toward the river. In O. G. Howland's words:

> One of the crew came in hatless, another shirtless, a third without his pants, and a hole burned in the posterior portion of his drawers; another with nothing but drawers and shirt, and still another had to pull off his handkerchief from his neck, which was all ablaze. With the loss of his eyelashes and brows, and a favorite moustache, and scorching of his ears, no other harm was done. One of the party had gathered up our mess kit and started hastily for the boat, but the smoke and heat was so blinding that in his attempt to spring from the shore to the boat he lost his footing and fell, mess kit and all, in about ten feet of water, that put him out (I mean the fire in his clothes), and he crawled over the side of the boat as she was being pushed off, not worse, but better, if possible, for his ducking. Our mess kit, however was lost ....

Finally, on June 18th, they reached the mouth of the Yampa River — out of the grip of Lodore's rapids, tired, but content to have behind them water that Bradley called "the worst we shall ever meet." Wishful thinking, perhaps.

At this wide, straight-walled meeting place of the Green and Yampa Rivers (later to be named Echo Park), the men found they could receive up to ten echos from a loud, sharp yell. Fishing was so good that when Bradley had lines broken by the huge squawfish, he braided lines together to land one of the big fish. Climbing was rewarding too, particularly in the view from the top of Steamboat Rock, 700 feet above the center of a horseshoe bend in the river. Two days were spent in Echo Park.

For Powell and his men, the canyons that led out of the Uintas were not nearly so difficult as those that led in. Whirlpool and Split Mountain Canyons have many rapids, but they are not the huge boulder and narrow channel type. Here the rapids are merely shallow and rocky — not dangerous to an alert boatman. Nevertheless, the cautious Powell preferred to portage and line a number of these rapids rather than risk another accident.

On June 26th, they suddenly emerged from Split Mountain Canyon and found themselves in the broad Uinta Valley, where the river meanders lazily between banks lined with cottonwoods and brush. Waterfowl were seen frequently, as were the "meanest pest that pesters man — mosquitoes." While making notes in his journal, Bradley complained that the "musical

One-armed Powell climbed the 700-foot Steamboat Rock in Echo Park at the junction of the Green and Yampa. Repairs, scientific work, and good fishing kept the group here for two days.

Echo Rock

little mosquitoes bite so badly that I can write no longer."

The mouth of the Uinta River, which comes in from the west, was important to Powell for a number of reasons. First, it was to be the last time in the journey that expedition members would have contact with an established village, which, in this case was the Uinta Indian Agency 30 miles up the Uinta River. Second, Frank Goodman decided to leave the expedition. Goodman had never adjusted well to the rigors of boat handling and his mishap in Lodore Canyon had cost him most of his personal belongings. The other men were not sorry to see him leave. Third, although no one knew it at the time, the Uinta River marked the limit of previous voyages down the Green River; from now on the Powell Expedition would indeed be the "first" on the river.

Major Powell waited at the junction a couple of days for Howland to finish a map, for Bradley to correct his field notes, and for the men to write letters to home. Then on July 2nd, he was off for the Agency, taking Hawkins and Goodman with him. Walter Powell and Andy Hall had already gone ahead. For the rest of the men it meant a boring Independence Day spent in the cottonwoods on a muddy river bank. On July 5th, the travelers returned, bringing with them 300 pounds of flour that the Major had traded for. Preparations were made for an early start in the morning. Bradley wrote that he "was exceedingly glad to get away for I want to keep going."

They passed the mouth of the White River, where on a river island they spotted a small garden. Helping themselves, they found that most of the vegetables were still immature, so they asked Andy to cook the green tops. All who ate them became sick and had attacks of vomiting, which fortunately, made the men feel both better and wiser.

By this time the men were through with the education phase of their journey and were prepared for the severe trials that lay ahead.

III
INTO DESERT CANYONS

From the Uinta Valley they saw to the south a low range of hills, though which the Green River entered by way of a steep-walled cleft through yellow sandstones. Powell was to name this "Desolation Canyon" for the barren aspect of its arid cliffs. When in the canyon they found that occasional hot winds roared through the gorge both in daytime and at night. As one man wrote, "We are quite careful now of our provisions as the hot blasts that sweep through these rocky gorges admonish us that a walk out to civilization is almost certain death, so better go a little slow and safe."

Since today's river runners have fine maps and know exactly where they are at all times, there is no need to obtain their bearings from high points. But the 1869 expedition was making the first map, and Powell, Howland, and others frequently had to scale the cliffs to obtain the lay of the land.

On one of these climbs, Major Powell, (who was missing his right arm), became trapped on a precarious cliff where he could go neither up nor down. When he called for help George Bradley made his way to a point above the Major, but he found that he was too high to reach Powell's hand. Bradley had no rope, but when he could not find an adequate stick, he thought of an ingenious expedient. Quickly he took off his trousers, lowered them down to Powell, who with one motion released his handhold and grasped the pant legs. Bradley pulled him to safety. In his journal, Bradley modestly stated that "my drawers . . . made an excellent substitute for rope and with that assistance he got up safe."

Rough water was frequently encountered in Desolation, and the Major often ordered a portage or lining procedure. Perhaps, as the men commented, the Major was cautious to a fault; most of the rapids could have been run without undue risk. Such over-caution was to cost the expedition dearly in terms of time, in terms of provisions consumed, and in terms of frayed nerves and general exhaustion.

When Desolation Canyon finally ended, another gorge, equally desolate, lined their route. This was named "Coal Canyon" by the Major, for the strata of coal seen along the cliffs. Now it is called Gray Canyon.

In spite of the bleak cliffs, the spirits of the men were still high. Bradley commented that:

Andy is singing for his own amusement and my edification a song that will no doubt some day rank with "America" and other national anthems. All I can make out as he tears it out with a

Light House Rock in Desolation Canyon.

Gunnison's Butte at the foot of Gray Cañon. (2,700 feet high.)

voice like a crosscut saw is the chorus: "When he put his arm around her she **bustified** like a forty pounder, look away, look away, look away in Dixie's land."

On July 11th the Major had to run a rapid and was swamped. He tried to land above the rapid, but the earlier loss of an oar handicapped his crew in their efforts to reach shore. A wave rebounding from a rock filled the **Emma Dean**. Then another wave turned the boat over, throwing its occupants into the water. But Powell and his men were wearing their inflatable life preservers and they suffered no harm. Although the boat was caught, they lost two rifles, a barometer, and some blankets. The rest of the day was spent drying out and finding suitable trees from which to make new oars.

Heat waves were rising from hot sunlit sands when the three boats emerged from Coal Canyon into the broad valley where Green River, Utah, now stands. In Powell's day the site was called Gunnison's Crossing, named for Capt. John W. Gunnison, who surveyed a railroad route through here in 1853. Earlier it had been the spot where the Old Spanish Trail (from Santa Fe to Los Angeles), crossed the Green River.

In 1869 no white men lived in this valley, and its desolate appearance suggested to Jack Sumner that it would remain forever uninhabited. "The upland is burned to death and on the river there are a few cottonwood trees, but not large enough for any purpose but fuel," he commented in his journal.

Beyond the valley they entered yet another canyon, although the walls seemed to be perpendicular on one side at a time. Red sandstone cliffs shimmered under the burning July sun, while on the placid, winding, muddy river, the three wooden boats floated slowly downstream. Bradley's imagery is unsurpassed.

> The whole country is inconceivably desolate, as we float along on a muddy stream walled in by huge sandstone bluffs that echo back the slightest sound. Hardly a bird save the ill-omened raven or an occasional eagle screaming over us; one feels a sense of loneliness as he looks on the little party, only three boats and nine men, hundreds of miles from civilization, bound on an errand the issue of which everybody declares must be disastrous.

But little actual pessimism was evident. Hadn't they already endured some of the worst that the river was likely to offer? That the river had only begun to show its fury was soon evident.

Narrow Canyon, leading west to Dirty Devil River. The Henry Mountains, last major mountain range to be discovered in the U.S., are in the distance.

IV
ON THE COLORADO

"Hurra! Hurra! Hurra!" was the shout when they rounded a bend to see the Grand River joining its waters to that of the Green to form the Colorado River. At this famous, but never-before-seen spot, cliffs towered above each river bank, leaving only narrow strips of willow-covered beach at the water's edge. Since they had been told that the Grand was a mountain stream, they expected to see a clear water torrent flowing into the Green. Instead, the Grand looked very much like the Green — same muddy water, same size, same placid flow.

For four days they remained in camp at a point between the Green and the Grand. Powell had previously planned to remain here much longer, in fact until the solar eclipse on August 7th, but an investigation of their supplies revealed the shocking fact that 200 pounds of flour had spoiled and would have to be thrown away. Their rations cut to a two-month supply, Powell reluctantly decided to proceed on down the river as soon as he could make a hasty analysis of the topography.

Powell's climbs out of the canyon at the confluence gave him close views of some of the most unusual formations in the land. Nearby was the Land of Standing Rocks, where 100-foot sandstone spires stand like a forest of weird totem poles. To the east were the pinnacles and canyons of the Needles, while more distant were the Sierra La Sal, bearing small traces of snow. Strange was the world that lay before him.

On July 21st, they were off again, at last floating on the Colorado itself, in the face of what dangers they knew not. Only four miles of calm water ensued before they reached the first rapid in what Powell was to name Cataract Canyon, for the almost unending succession of tumbling white water. The first day they made only 8½ miles, including four portages and a capsizing of the **Emma Dean**. Bradley succiently stated, "I conclude the Colorado is not a very easy stream to navigate."

They reached a bad rapid late in the day and decided to camp where they were and to make the necessary portage in the morning. Although Cataract Canyon has many fine beaches and places to camp, they found that where they were was not one of the ideal spots. Consequently, for places to sleep they filled sand in between the boulders and crawled in for an uncomfortable night.

Lining boats and portaging became almost incessant in the days that followed. Rapid followed rapid so closely that the men had little rest, and they had almost none of the exhilaration enjoyed by modern river runners. The steady descent of the river caused Bradley to write:

We know that we have got about 2,500 ft. to fall yet before we reach Fort Mohavie (sic) and if it comes all in the first hundred miles we shan't be dreading rapids afterwards for if it should continue at this rate much more than a hundred miles we should have to go the rest of the way **up hill,** which is **not often the case with rivers.**

Rapids don't interest me much unless we can run them. That I like, but portage don't agree with my constitution.

On July 28th, the river course veered to the west and the rapids ceased. Cataract Canyon lay behind them, while anead was a range of basaltic mountains never before mapped by man. They called them simply the "Unknown" Mountains, but later Powell named them the Henry Mountains in honor of Joseph Henry of the Smithsonian Institution.

Canyon walls dropped lower as they neared a small stream that entered from the north. One man hollered to Bill Dunn, who was standing beside the creek, "Is it a trout stream?" Dunn looked disgustedly at the brown, odoriferous liquid and shouted, "No, it's a dirty devil." Powell thought the name so appropriate that he named it the Dirty Devil River.

A new canyon lay before them, a new canyon formed from colorful red sandstone eroded into buttes and alcoves, into sheer cliffs and patterned grottos. Powell named it Glen Canyon, for the many tree-filled glens found at the mouths of side creeks. But where today's tourists find the magnificence and splendor of Lake Powell, the men of the 1869 expedition could see little but sterility and lack of farm land. Such is the difference in outlook between a frontiersman and his counterpart of today.

The mouth of the San Juan River, which was one of the few geographical points that they could expect to see, was reached on the 31st of July. Two miles below the San Juan, upon investigating a small side canyon, they discovered a deep alcove, where the canyon walls nearly met at the top, and which contained deep green ferns, a dripping waterfall, and a quiet pool at the far end. This alcove they named "Music Temple," for the song that Capt. Walter Powell sang within its shadowed walls. Some of the men, using knives, incised their names into the soft sandstone wall.

After a night in Music Temple, Powell ascended the naked sandstone wall to a point where he had a commanding view of the surrounding land. Within this world of rock he located the Henry Mountains, while nearby, to the south, rose the steep, eroded face of Navajo Mountain. To the north was the Kaiparowitz Plateau and its outliers. But to Powell most formations had no names — and no history.

An historic point, however, was soon reached when they came upon the Crossing of the Fathers, the site where two Franciscan priests and their entourage had crossed the river in 1776. Actually, Fray Escalante and Fray Dominguez had simply used an old Indian ford, a ford that was still in use during 1869. Powell noted the hoofprints on each side of the river and was thereby able to locate the crossing precisely.

On August 4th they passed what would be the site of Glen Canyon Dam, but the Major only remarked that "the walls grow higher, and the canyon much narrower." Although they were on the lookout for the mouth of the Paria, they camped within a few hundred yards of it that night without recognizing it. While at the Paria, Sumner thought he was at the Crossing of the Fathers and wrote that it was "desolate enough for a lovesick poet." Little did they anticipate that the spot of land where they camped would in a few years be known as Lee's Ferry, where much future history would take place.

Cataract Cañon

LAND OF THE STANDING ROCKS.

V
A CANYON OF POLISHED STONE

With some feeling of anxiety, we enter a new cañon this morning. We have learned to closely observe the texture of the rock. In softer strata, we have a quiet river; in harder, we find rapids and falls. Below us are the limestones and hard sandstones, which we found in Cataract Canyon. This bodes toil and danger.

With these words, Powell describes his thoughts upon entering what he was to name Marble Canyon. Incised through a valley floor composed of hard Kaibab limestone, Marble Canyon grows deeper as it descends, until, at its lower end, it is about 3,000 feet deep. No wonder that Powell's men thought they were riding a river into the bowels of the earth!

As Powell had anticipated, the rapids soon began. Although their six days in Glen Canyon might have seemed slow and boring to this rough-and-tumble crew, it was like Heaven compared to the heavy labor and near exhaustion that awaited them in Marble Canyon. In spite of their daily complaints about rapids and portages, by the time they passed Lee's Ferry, the Powell crews of 1869 were pretty fair boatmen. No more did they blunder into rapids and smash boats as they did on the Green River; they now recognized dangers instantly and could quite deftly handle the boats with the oars. It is too bad that no one was there to tell them just how good they were, for their morale was beginning to sag. As for ex-Army Sergeant Bradley:

Thank God the trip is nearly ended for it is no place for a man in my circumstances but it will let me out of the Army, and for that I would almost agree to explore the river Styx.

Even their increased skill, however, was not enough to keep the boats off the rocks entirely. The **Maid of the Canyon** struck hard against a submerged boulder, knocking a hole in the hull. The damage was repairable, but as Bradley commented, "Have been in camp all day repairing boats, for constant banging against rocks has begun to tell sadly on them and they are growing old faster if possible than we are."

Their meager supply of clothing too was feeling the heavy strain of weather and abrasive rocks: "We begin to be a ragged looking set for our clothing is wearing out with such rough labor and we wear scarce enough to cover our nakedness."

Major Powell knew well in advance of his trip that a solar eclipse was due on August 7th, and that the eclipse should be visible from his location. Since the time that the moon would be-

Noon-day rest in Marble Cañon.

gin to block the sun's rays was listed in the almanac, Powell could use his observation of the eclipse as a check on his chronometers, and hence could accurately determine his longitude. The Major halted the expedition early in the day, and he and his brother climbed the cliffs to a good vantage point. But August is a rainy month in the Southwest; the afternoon clouds rolled in; the sun was obscured. Quite disappointed, the Powell brothers tried to return to camp. Rain began to fall in torrents, and the night became too black to descend the rocky cliffs of Marble Canyon. They huddled along a ledge, trying to avoid the worst of the rain, while they waited for the gray light of dawn that would allow them to return to camp.

In spite of discouragements, Marble Canyon had some scenic allurements that appealed to the tired, hungry travelers. One day they rounded a bend to discover a clear spring flowing from the cliff, the water discoursing down through a lush growth of ferns, flowers, and trees. Poetically, Powell described it:

> Every eye is engaged, everyone wonders. On coming nearer, we find fountains bursting from the rock, high overhead, and the spray in the sunshine forms the gems that bedeck the wall. The rocks below the fountain are covered with mosses, and ferns, and many beautiful flowering plants. We name it Vasey's Paradise, in honor of the botanist who traveled with us last year.

Bradley stated that it "was the prettiest sight of the whole trip." Sumner agreed to an extent by writing, "The white water over the blue marble made a pretty show. I would not advise anybody to go there to see it."

They also camped in the huge river-dug ampitheater now known as Redwall Cavern. Their high appreciation of its dry sandy floor stemmed from the fact that it was still raining almost every afternoon and evening.

On August 10th, at two in the afternoon, their southwestern course suddenly veered straight west, and they came upon the mouth of the Little Colorado River — another of the landmarks they searched for. Here they planned to stop two or three days to determine their latitude and longitude before they began the descent down the unknown waters of the Grand Canyon. Three days might have seemed short to the scientifically curious Major, but it was harsh punishment to the tired and hungry men. Bradley commented in his diary:

Our camp is under the shelving edge of a cliff on the south side of the Chiquito (Little Colorado) and is protected from both sun and rain by overhanging rocks though it is filthy with dust and alive with insects. If this is a specimen of Arrazona (sic)- a very little of it will do for me. If Major does not do something soon I fear the consequences but he is contented and seems to think that biscuits made of sour and musty flour and a few dried apples is ample to sustain a laboring man. If he can only study geology he will be happy without food or shelter but the rest of us are not afflicted with it to an alarming extent.

View from Camp at the Mouth of the Little Colorado

Compare this sketch of Marble Canyon with the aerial photo opposite. Lee's Ferry and Paria River are near the bottom. The general outline and character of the land are remarkably similar.

Marble Canyon winds southward from Lee's Ferry which was opened in 1872. Powell camped here on August 5, 1869 not knowing the future importance of the place. This is the present starting point for Grand Canyon boat expeditions.

VI
INTO THE GRAND CANYON

While it was true that Major Powell thought about science, he was still mindful of the necessity of survival. He knew that in the Grand Canyon they faced the greatest challenge of the entire trip. In his diary he reflected on their relative helplessness in the face of raw, brutal — but magnificent — Nature:

> We are three-quarters of a mile in the depths of the earth, and the great river shrinks into insignificance, as it dashes its angry waves against the walls and cliffs, that rise to the world above; they are but puny ripples, and we but pigmies, running up and down the sands, or lost among the boulders.

> We have an unknown distance yet to run; an unknown river yet to explore. What falls there are, we know not; what rocks beset the channel, we know not; what walls rise over the river, we know not.

Soon after they left the Little Colorado they encountered along the walls of the river a new rock formation instantly recognized as a threat to their success.

> Heretofore, hard rocks have given us bad river; soft rocks, smooth water; and a series of rocks harder than any we have experienced sets in. The river enters the granite! . . . As we proceed, the granite rises higher, until nearly a thousand feet of the lower part of the walls are composed of this rock.

Bad rapids did begin again, and in the estimate of every man, they exceeded in ferocity anything they had believed possible. "Wildest day of the trip so far!" exclaimed one man. "A perfect hell of waves," described another. As they did before, some of the rapids were run, many were lined, and a few were actually portaged, with the heavy boats being carried over the boulders on the shore. Even the campsites were sometimes uncomfortable, as Bradley described the site for the night of August 14th:

> We have but poor accommodations for sleeping tonight. No two except Major and Jack can find space wide enough to make a double bed and if they don't lie still we shall "hear something drop" and find one of them in the river before morning . . . The rest are tucked around like eve swallows wherever the cliff offers sufficient space.

THE GRAND CANYON.

In spite of their agonies and hard work, they did have certain pleasures. One of the most pleasant camps, for instance, was reached on August 15th. A large weeping willow gave them shade as they relaxed beside a crystal brook that flowed down from the north. At first the Major called it Silver Creek, but a few months after the journey he bestowed on it the name it bears today — Bright Angel Creek.

Modern river runners would realize that by the time Powell and his men reached Bright Angel Creek, they had already passed some of Grand Canyon's major rapids, including Hance, Sockdologer and Grapevine. Many bad rapids, however, still lay further downstream. In their half-starved and exhausted condition, the men would have to cope with Granite, Hermit, Deubendorff and Lava Falls Rapids.

Powell gave names to none of these rapids on his 1869 expedition. It is reasonable to suppose that since he never envisioned an era of river running for recreation, he wasted neither thought nor time on geographical features, such as rapids, that almost no one would ever see. Still, it is strange that he named at least four rapids on the Green River, but he did not bother to name the much larger ones on the Colorado.

So the men rested under the weeping willow while Powell climbed the slope, searching for suitable pine trees from which to make oars.

When the chores and the rest stop were over, on August 17th, they started off again down the river. Shortly before leaving the cook spread the meager rations out on a boat to dry. The boat began swinging with an eddy current and the rope that held the boat to the bank scraped across the rations and knocked the only box of baking soda into the river. It was a small matter, but to the men who would henceforth eat unleavened bread, the loss of the soda was a minor catastrophe.

As they left Bright Angel Creek Powell wrote:

> The stream is still wild and rapid, and rolls through a narrow channel. We make but slow progress, often landing against a wall, and climbing around some point, where we can see the river below. Although very anxious to advance, we are determined to run with great caution, lest, by another accident, we lose all our supplies. How precious that little flour has become! We divide it among the boats, and carefully store it away, so that it can be lost only by the loss of the boat itself.

Although their rations were no burden, the boats weighed as much as ever (perhaps more, considering that they had slowly

The Grand Canyon looking West from Toroweap.

soaked up water for three months). A portage, no matter how short, was a grueling undertaking. Yet on their first day out from Bright Angel Creek, they had to portage three times. Above them still towered the walls of the granite gorge, which allowed them only brief views of the distant and lofty upper cliffs of the Grand Canyon.

The next day an unpleasant thunderstorm allowed Bradley a chance to wax lyric in prose:

> We had to fasten our boats to the rocks and seek shelter from the wind behind bowlders (sic). The rain poared (sic) down in torrents and the thunder-peals echoed through the canon from crag to crag making wild music for the lightning to dance to. After a shower it is grand to see the cascades leap from the cliffs and turn to vapor before they reach the rocks below. There are thousands of them of all sizes, pure and white as molten silver.

Wild, close escapes became commonplace. One early afternoon the small boat, the **Emma Dean,** was quickly swamped in a rapid. To the rear, the men in the other boats instantly saw the trouble, but while rushing to give help, Bradley's boat struck a rock with great force. Fortunately, the heavy wooden cutwater took the impact and the craft was not damaged. By the time they reached the **Emma Dean,** the Major and his crew were all in the water holding up the boat.

On August 21st, they passed beyond the granite walls, giving indication that smoother water probably lay ahead. In Major Powell's words:

> Ten miles in less than half a day, and limestone walls below. Good cheer returns; we forget the storms, and the gloom, and cloud-covered canons, and the black granite, and the raging river, and push our boats from shore in great glee.

Rapids had by no means ceased entirely, but the prospect was encouraging. At the very least the geology and topography became more interesting. They stopped at Tapeats Creek, which flows in from the north, and Powell walked up this rushing, clear stream, then up a tributary, Thunder River. Often he had to wade up to his neck. Finally, he came to the now-famous Thunder Falls, which the Major estimated to be 150 feet high.

Below Tapeats Creek, they discovered the magnificent waterfall now known as Deer Creek Falls, where a clear stream pours through a narrow crevice, dropping 100 feet to a deep pool.

The Grand Canyon viewed East from Toroweap.

Through this area the canyon walls were becoming farther apart, but they still rose over 3,000 feet above them.

One day they came to an area abounding in great, broken pieces of solidified lava standing like sentinels from the river up the north wall. The geologist in Powell saw that in some recent geologic time, a canyon rim volcano had poured great amounts of molten rock into Grand Canyon, forming at this point a natural dam. Since then the relentless river had cut away the lava dam, but it had left a rapid of horrendous proportions.

In writing about the lava flow that had caused this feature, Powell paints a graphic picture:

> What a conflict of water and fire there must have been here! Just imagine a river of molten rock, running down into a river of melted snow. What a seething and boiling of the waters; what clouds of steam rolled into the heavens!

The rapid at the lower end of the lava flow is now known as Lava Falls (or Vulcan Rapid). Powell of course bypassed the rapid by means of a portage.

Vulcan Rapid, or Lava Falls

38

VII
ESCAPE FROM PRISON

That night the cook opened the last sack of flour. The moment was an ominous one. Significant of their state of mind was Powell's diary reference, on August 26th, to the Grand Canyon as a "prison."

August 27th was a fateful day for the expedition. The river took a sharp turn to the south, taking the men into lower geological formations, and toward the hateful granite. A turn or two to the west briefly buoyed the mens' spirits but the canyon again cut to the south. By mid-morning the black granite walls began to appear.

> About nine o'clock we come to the dreaded rock. It is with no little misgiving that we see the river enter these black, hard walls. At its very entrance we have to make a portage; then we have to let down with lines past some ugly rocks. Then we run a mile or two farther, and then the rapids below can be seen.

Just before noon they came to a place in the river that seemed impassable. A furious rapid in a narrow chute roared over huge boulders, and walls of water crashed against the cliffs. Worst of all, there was no place to walk along the rapid, so the boats could neither be lined nor portaged around. If the men were to reach the river below, they would have to run this rapid.

In examining their situation, they noticed a long, deep side canyon coming in from the north. To these canyon-imprisoned men, facing what appeared to be a choice of death by starvation or death by drowning, this side canyon looked like it might offer a way out. Oramel Howland assumed the leadership of a small faction rapidly becoming convinced that the voyage should be abandoned, and that the men should strike out overland for St. George, the nearest town.

That afternoon Major Powell examined the rapid as best he could, and returning to camp, announced that they would run the rapid in the morning.

Powell's decision was not acceptable to Oramel Howland. That evening after dinner, Howland told Powell that three men — Oramel and Seneca Howland and Bill Dunn — had made up their minds to climb out. When Oramel implored Powell to join them Powell wavered. Was he being foolishly persistent? Did they indeed face certain death in Grand Canyon?

By dead-reckoning, Powell estimated that they were about 45 airline miles from the mouth of the Virgin River, where they knew white men were living. He estimated that St. George must be about 75 miles away. Powell knew enough of the coun-

try to be sure that the 75 miles were mostly harsh, unfriendly desert.

Late that night Powell awoke Oramel Howland to reaffirm his decision that they continue on the river, since this appeared to him to be the safer course. Then, one by one, Powell went the rounds of the sleeping men, waking each to find out how he stood on the issue. Five men agreed to go with Powell. The Howland brothers and Bill Dunn were determined to walk out.

The words are Major Powell's:

> **August 28** — At last daylight comes, and we have breakfast, without a word being said about the future. The meal is as solemn as a funeral. After breakfast I ask the three men if they still think it best to leave us. The elder Howland thinks it is, and Dunn agrees with him. The younger Howland (Seneca) tries to persuade them to go on with the party, failing in which, he decides to go with his brother.

Having three less crewmen, Powell decided to abandon one of the boats, the **Emma Dean,** the smallest of the three. He gave to the Howlands and Dunn a duplicate set of the expedition records, plus two rifles and a shotgun. Powell wrote a letter to his wife and gave it to Oramel Howland.

> Some tears are shed; it is rather a solemn parting; each party thinks the other is taking the dangerous course.

To lighten the boats for the big rapid coming up, Powell even abandoned much of his scientific gear and his collections of mineral and fossils.

Bill Dunn and the Howland brothers stood on a low cliff watching as the two boats were swept into the churning rapid. They were seen to graze a rock, then pass into a hole where the open compartments filled with water. Oarsmen pulled hard to avoid a dangerous rock on the right. Then the boats passed around a bend beyond the sight of the men on the cliff.

At the foot of the rapid, Powell had his men fire their guns as a signal that they had come through safely. Here they waited two hours, hoping that the Howlands and Dunn would follow in the **Emma Dean.** Finally, they continued on down the river.

The big rapid that caused the party to split is now known as Separation Rapid, while the side canyon to the north is Separation Canyon. A plaque on the cliff marks the fateful spot.

That afternoon Bradley got out of a scrape that won him the admiration of the others. When lining a boat they had to ascend a low cliff to get around the rapid, but they found that their rope was too short. Bradley, who was aboard the boat to steady it, found himself in a maelstrom of white water and swinging from rock to rock while the men on the cliff went back for more rope. Sensing that his boat would be smashed if he remained where he was, Bradley rushed to the bow to cut the rope. But as he leaned over, the stem post on the boat broke away and flew thirty feet into the air. With perfect composure, Bradley grabbed the oars, pulling first on one side and then on the other to avoid the rocks, and within a few seconds was at the foot of the rapid.

After the hectic day of August 28th, their situation began to improve. On the 29th they ran out of the granite, and about noon the terrain suddenly fell back. They had reached Grand Wash at the foot of the Grand Canyon. Powell writes:

> The relief from danger, and the joy of success, are great. . . How beautiful the sky; how bright the sunshine, what "floods of delirious music" pour from the throats of birds; how sweet the fragrance of earth, tree and blossom!

The next day they found some timid Indians on the bank, but Powell could not make himself understood. Later they again saw an Indian, only this time there were three white men beside him, hauling in a seine. These were Mormons, a Mr. Asa and his two sons, who knew immediately that the boating party was Major Powell and his expedition, for they had been instructed to be on the lookout for them — or for their remains.

Sumner wrote that the kindly Mormons cooked them a fish dinner, and that the guests "laid our dignified manner aside and assumed the manner of so many hogs. Ate as long as we could and went to sleep to wake up hungry."

On August 31, 1869, the expedition ended, when Bishop Leithhead and some Mormons arrived from the little town of St. Thomas. On the Bishop's wagon were many fine, juicy watermelons, which Powell and his brother ate as they rode on their way back to civilization. The other men, Sumner, Bradley, Hawkins, and Hall took their share of the watermelons plus some rations given to them by the Mormons, and set off down the river for Ft. Yuma, which they reached without incident.

VIII
THE AFTERMATH

Oramel Howland, Seneca Howland, and Bill Dunn never showed up at the Mormon settlements. On Sept. 7, 1869, the **Deseret Evening News** of Salt Lake City reported an unconfirmed story that the three men had been killed by Shivwits Indians, but no one knew for sure.

A little over a year later, in September, 1870, Powell returned, not to the river, but to the Arizona Strip country north of the Grand Canyon. He was led to this barren land by Jacob Hamblin, the famous Mormon Indian guide and "Apostle to the Lamanites." Powell had in mind not only finding the solution to the Howland-Dunn mystery, but also learning something about the habits and mores of the primitive Shivwits Indians.

In a solemn evening conference Hamblin spoke to the Shivwits leaders seated around a campfire. Hamblin told them about Powell's expedition down the river and about the three men who had climbed out. The Indians thereupon freely admitted that they had killed the three men, but said that they believed them to be miners who had abused their women. The Howland brothers and Dunn were killed, they said, when the white men were obtaining water from a spring. Had they known the truth about the men, the Indians said, they would have aided the white men instead of killing them.

Jacob Hamblin and Powell in council with Shivwits Indians.

Later in 1870 Powell followed Hamblin on a peace mission to the Navajos at Ft. Defiance. On that trip they became the first men to use the site of Lee's Ferry as a boat crossing.

Because the 1869 river expedition had been rushed, and since the personnel aboard were generally untrained in scientific observation techniques, Powell concluded that another river expedition would be necessary. The second expedition therefore set forth from Green River, Wyoming, on May 22, 1871. This group reached Lee's Ferry in late October and left the river for the winter. They returned in August, 1872, and continued on down through Marble and Grand Canyons to the mouth of Kanab Creek, where the trip was abandoned.

In terms of sheer hardship, the 1871-72 expedition was quite comparable to that of 1869. But no one can deny that the first expedition faced what is possibly the greatest danger of them all — the unknown.

In later years, Powell's achievements in government and in science certainly outranked in importance his adventure of exploring the Colorado River. He led one of the great topographic surveys of the West, he was instrumental in consolidating the surveys into the U.S. Geological Survey, and he was the second Director of the U.S.G.S. He pointed the way in Indian research and was first director of the Bureau of American Ethnology. Feeling that pioneer settlement of the West was based, not upon reality, but upon an archaic pattern established in more humid climates, he campaigned for a new Homestead law that would recognize the importance of water sources. Eventually he lost this battle, but his ideas were incorporated in the 1902 legislation to establish the Bureau of Reclamation.

When Powell died at the age of 68 on August 23, 1902, his death was considered a great loss to the world of science. Scientific leaders throughout America and the world paid tribute to this man who had seen the nature of the West so clearly.

More than anything else, the 1869 river expedition provided a measure of the man that was to make such a heavy mark upon the character of American science and upon our pattern of Western development. As a one-armed cripple, he had faced the unknown, he had emerged triumphant. His courage, his wisdom, and his indomitable spirit were displayed to the world. Powell's pioneer voyage will ever rank with the greatest stories of pioneer heroism that American history has produced.